Pony Holiday Book

Lucy Popescu

Pony Holiday Book

Illustrated by Peter Clark

A DRAGON BOOK

GRANADA

London Toronto Sydney New York

Published by Granada Publishing Limited in 1982

ISBN 0 583 30420 6

Granada Publishing Limited
Frogmore, St Albans, Herts AL2 2NF
and
36 Golden Square, London W1R 4AH
866 United Nations Plaza, New York, NY 10017, USA
117 York Street, Sydney, NSW 2000, Australia
100 Skyway Avenue, Rexdale, Ontario, M9W 3A6, Canada
61 Beach Road, Auckland, New Zealand

Printed and bound in Great Britain
by Cox & Wyman Ltd, Reading

Phototypesetting by Grainger Typesetters Ltd,
Southend-on-Sea, Essex

Set in Baskerville

To friends and staff at Culford School, Bury St Edmunds, especially Fritz and Jonny who didn't believe this would happen.

Acknowledgements

I wish to thank the following:
Lord and Lady Fisher of Kilverstone Wildlife Park, Mr & Mrs Horton of The Decoy Riding and Livery Stables, Mr Leslie Mills of the Dedham Heavy Horse Centre, Courages Shire Centre, Charlotte Popescu for her Crosswords, Glenda Spooner of *Ponies of Britain* and my mother, Christine Pullein-Thompson, for help and advice.

CONTENTS

Trekking 11
Riding lessons 14
Selected trekking centres 16

Put your pony first 23
A cautionary poem 23
Leaving your pony behind . . . 24
. . . and taking him with you 26
Coping with crises on holiday 27
What did Sarah do wrong? 30

Places to visit 37
Sea poem 37
How to Find Out About Places 38
A visit to the Royal Mews 39
Heavy Horse Centres 42
Kilverstone Wildlife Park 46
Native Pony Breeds Centres 50
Places to visit 55
Trekking centres 57
Annual events 59

Rainy day miscellany 63
Limerick 63
Pony feats 64
Puzzles and quizzes 66
Answers 81
Books to read 89

Foreword

I know from past experiences that there are times when a holiday becomes boring or tiring. This book is intended to fill those moments when you are at a dead loss as to what to do and what to see. It tells you how to make the best of such times. There are quizzes and games to do, crosswords by my sister Charlotte and places to visit all over Great Britain. There are also poems, word searches, and bits and pieces for you to do, and read – all connected with horses of course.

Happy reading for those among us who fit into the category of horse and holiday lovers.

Lucy Popescu

TREKKING

Trekking usually starts during April and continues until the end of September – or into October in the Lake District and Scotland. There are approved centres in Scotland, the Lake District, in the Fells and Dales, the North of England, and in Wales. There are also approved centres in the USA, Spain and Ireland. Trekking is suitable for novice riders because most of it takes place at a walk. Treks start at around ten in the morning and finish in time for tea with packed lunches provided; so you do need to be fairly fit. Costs vary from £50 to £200 a week. Anything cheaper should be avoided as such

centres tend to underfeed their ponies or their visitors! If you are heavy go to a centre which provides a heavier type of pony, such as the Dale Highland or Welsh Cob. A Dartmoor or Welsh pony should not be expected to carry anyone over nine stone all day long. And never trek on a pony less than four years old.

Post trekking, that is moving on from place to place, sometimes staying away from the actual centre for several days, is not suitable for novice riders.

If you want tuition, it is probably better to book a riding holiday rather than a trekking one. These vary enormously and may provide riding for your parents as well as golf, tennis, fishing and sailing, so if your family plans to go with you, this might be the better choice.

Finally, always check everything before you set out on a trek. Your pony's saddle should fit. If it is pressing firmly on his withers it does not fit. His hoofs should be well shod and look tidy. If they are full of protruding nails and clink at every step, he is in no condition to trek. His bit should look comfortable in his mouth, and not be hanging among his teeth. Finally he should appear well. If his ribs protrude and he staggers under your weight, cancel your ride. You will not be popular, but it will be the right decision.

You may decide to take some riding lessons before embarking on a trekking holiday. Here are some helpful 'dos and don'ts' on how to behave at a riding school.

Riding Lessons

Dos

Take only one other person with you. Your first ride may be a great moment, but a crowd of people, following you from loosebox to loosebox making inane remarks, will get in the way.

Dress in jodhpurs if you have any.

Be polite. It is better to say nothing if you have nothing nice to say.

Be truthful about your weight and your experience. If you don't like galloping say so.

Be prepared to pay a reasonable sum for a lesson. Less then £2 an hour may mean underfed and overworked ponies.

Help saddle and unsaddle your pony.

Wear a hard hat with a safety harness or elastic under your chin.

Say 'thank you' before you leave.

Don'ts

Don't wear unsuitable shoes – shoes with high heels or ridges can become jammed in stirrups. Shoes without heels can slip right through. Wellington boots were not designed for riding!

Don't wear jewellery. It looks wrong and is liable to break and become lost. Have you ever tried looking for a bracelet in a fifty acre field?

Don't wear decorative slides and hair combs under your riding hat. They can stick into your head if you fall off.

Don't boast about your riding even if you have won a hundred rosettes and trophies.

Don't take your dog with you. He may lose his head and chase beginners, with disastrous results.

Don't take your little sister or a friend, on a bicycle, either. They will be easily knocked over and are out of place in the middle of a ride.

Don't criticise your pony or expect to be given the best pony in the stable at your first appearance.

Don't complain if there are no heart warming gallops. If you are good enough, your instructor may move you into a better and faster ride next time, but grumbling won't help.

Don't throw your weight about and expect constant individual attention unless you have paid for an individual lesson – when you should get it anyway.

Don't ride at a stable where the ponies are thin, underfed and under age (less than four) or where the tack does not fit. Complain to someone – your local council can refuse a licence or the British Horse Society may be able to help. In a really bad case go to the RSPCA. A pony lover should never finance cruelty.

Selected Trekking Centres

Scotland is a popular region of Britain for a trekking holiday. We visited briefly two trekking centres on the Isle of Mull. We went too early in the year to see them in action but the countryside is lovely and the horses friendly.

The **Erray Estate** has activity holidays, which means something for all of you. It is approved by everyone - The British Horse Society, The Scottish Trekking and Riding Association and Ponies of Britain. It offers trail riding for experienced riders only and riding and trekking for beginners and experienced riders between five and twenty-five years and weighing under 15 stone! The mounts are mostly Highland and experienced riders can canter and pop over the odd jump. Accommodation is provided in self-catering cottages built in 1719 but now modernized. They sleep four in two bedrooms. Dogs are welcome but must be kept under control.

If you are experienced, trailriding, that is cross-country riding, could be fun. Each day you will cover fifteen to twenty miles and then stay in different hotels around the island. There is also loch fishing, clay pigeon shooting, archery and Tobermory nine hole golf course is nearby. The address to write to for further details and up-to-date prices is on page 58.

The other trekking centre is at **Ormsaig, Ross of Mull.** Run by Penny Morgan, it seemed a very friendly establishment with mounts to suit all shapes and sizes. There are 2,000 acres to trek over with wonderful views of Iona, Coll and Tyree. There is no accommodation provided and it seems a pleasantly informal establishment. The riding costs less than at the Erray Estate. For further details see page 58.

The Decoy Trekking Centre — Beaulieu has a super outdoor school and facilities for studying for the British Horse Society's Assistant Instructor's Certificate, as well as a nice little cross-country course .

There is accommodation for twelve in bunks at the farmhouse, with a cooked breakfast, a sandwich lunch and hot evening meal provided. As on most riding holidays, you are allotted one horse or pony for your visit and expected to look after it . . . so the right clothes are important. I suggest the following plus other necessities:

warm sweater	riding boots
two pairs jeans	running shoes or sandals, etc
jodhpurs	Wellington boots
T-shirts, shirts, etc	plasters
socks - thin and thick	insect repellent
waterproof riding mackintosh or sensible anorak or huskie	antiseptic ointment
pyjamas	riding hat

Some establishments may organize a party, horse show or barbecue at the end of the week, so make enquiries and take suitable clothes for any such event. (A white shirt and a tie is often a must for a horse show.)

We found the Decoy Trekking Centre a very

friendly establishment with friendly staff and friendly horses and, because there is a stallion at stud here, plenty of foals about. All the treks are in the New Forest, so most of the horses are only shod in front. The tack room has certificates on the walls and smells of saddle soap and glistens with clean tack. Day treks start at ten in the morning and end around four in the afternoon. If you opt for a riding holiday, you will be given a lecture or lessons in the morning (horsey of course) and a trek most afternoons.

Beaulieu station is less than a mile away along a gravelled drive; otherwise there are the moors with mares and foals everywhere, endless heather and not much forest to be seen; Lyndhurst is within easy reach. So is Southampton Water. There are more than thirty horses here and the New Forest mares and foals are often around, so there is plenty of variety for the horse mad.

There are suitable mounts for anyone from midgets to fourteen stone fathers, so family outings on horseback are a distinct possibility. There are three instructors, and three working pupils, the proprietor, plus part-time staff, so you should find someone on your particular wavelength.

We did not try the food, nor the beds, but you are expected to bring your own bed linen (sheets rather than sleeping bags!). There is a pond stocked with fish for fishing enthusiasts but you must bring your own tackle.

A week's accommodation costs around £60 per person, more in August, and is likely to go up. This does not include riding.

A member of staff is allotted to look after unaccompanied children all day, so you can stay without parents. There are also plenty of camping sites, bed and breakfast places and hotels nearby, so you can always indulge in horsey activities while the rest of your family enjoys a more conventional holiday.

PUT YOUR PONY FIRST

A CAUTIONARY POEM

'He's dead, he's dead,'
They cried in vain,
Pulling at the leading rein.
'Why did we ever go away,
To sit upon the golden sands
And dig it with our busy hands?
We should have known he needed us,
To visit him each passing day
To check his water and his hay.
Instead we left him all alone;
And now he's dying, dying - died,
We have no pony left to ride.'

Anon.

Leaving Your Pony Behind . . .

Probably when you go away on holiday you won't be able to take your pony with you. However, you cannot leave him propped against a wall like a bicycle. He will need his water trough kept full, food of some sort or other and someone to keep an eye on him. Deserted ponies have a habit of getting out, of cutting themselves on wire, of developing a complaint which needs immediate treatment; you cannot simply walk out and forget your pony for two weeks.

If you are going to be away for a month or more have your pony's shoes removed and his feet rasped. Your blacksmith will do this and your pony will benefit enormously by going barefoot for a time. Find someone knowledgeable to visit him every day, make sure he has enough food - but not too much. A small field of less than an acre will soon grow bare particularly if there are several ponies in it, so do check that the grass will last the length of your holiday, or provide supplementary rations. Too much grass and no work may cause laminitis, so you may have to arrange for your pony to be exercised or taken in during the daytime. Winter feeding must be carefully organized, and a stabled pony will need mucking out and exercise. Rugs need adjusting too.

If you cannot make suitable arrangements for your pony at home, it may be better to send him to a professional establishment where he can be looked after. This may be expensive but could be less costly than vet's bills caused by tetanus, laminitis and the variety of other illnesses which occur when a pony is neglected. Your pony will enjoy the change of scene too.

Otherwise lend him to a really dependable friend, leaving a telephone number to ring should anything go wrong. And do remember when you return that if he has had a rest he may not be very fit and so ride him gently for a day or two.

. . . and Taking Him with You

Perhaps you will be fortunate enough to spend a holiday somewhere where your pony can come too. If the journey is a long one, make sure your pony has food and water while travelling. The trailer should be well bedded and the pony's legs should be protected by bandages. Take suitable rugs for any change in climate. Make sure he is well shod so that you do not need a blacksmith on arrival. Take a supply of feed with you unless you are certain it will be provided.

Once you are there, give him time to settle down. If he is sometimes difficult to catch, turn him out in a headcollar. Introduce him gradually to his equine hosts. Often it helps if the sexes are kept apart, ie, mares in one field, geldings in another.

When riding out, remember that your pony is likely to shy more and be nervous in a strange place, so ride accordingly. He may not like crossing bridges, railway crossings or galloping on sand – the sea may terrify him too – so give him time. Don't start labouring him with your whip, before he has had time to look around and collect his wits. He may never have seen the sea before, nor black faced sheep, nor peaty bogs. And if he won't approach them, he is only using his pony instincts. Leading him might help, because seeing you go first will give him confidence.

Watch his diet too. A change of grass or too much grass can cause laminitis especially if you don't exercise him for a few days.

Finally, if you live in a flat area, do not expect your pony to gallop up and down hills immediately. He will need to get fit to do this and he may need to improve his balance before he can cope with galloping downhill. Be sympathetic. Think how you would feel if you were suddenly asked to run up a mountain, full speed, with a weight on your back. If something does go wrong and you need a vet, the police or telephone operator will give you a number to ring without delay.

Coping with Crises on Holiday

If your pony disappears from his field search the immediate neighbourhood as quickly as possible; then telephone the police and any nearby riding establishment which he might have joined. Leave

someone to man your nearest telephone and continue searching, reporting back at frequent intervals. If you think he has been stolen, also ring the abattoirs giving a description.

If your pony develops colic summon a vet immediately. The local telephone exchange will give you the number of the nearest vet. Walk your pony up and down to prevent him from rolling while you await the vet's arrival.

If your pony loses a shoe ring the nearest riding school, saddlery or anyone else horsey for the address of a blacksmith. Be very polite. All blacksmiths are over-worked and most are choosey about whom they take on. So ultra politeness is necessary. If you ride him to a forge, keep on the grass verges as much as possible to save his unshod hoof wear and tear.

If you find a pony loose on the beach talk quietly to

him, catch him and look around for his owner. If one fails to appear put him somewhere safe while you inform the police and any nearby riding establishments.

If you find a pony locked in a field without food or water wait for at least half a day because he may have laminitis or be there for some reason (unless he is obviously dying of starvation, when no time should be wasted). Ring the RSPCA or the police. Give the pony's exact location. Be calm, cool and collected. If your parents are with you get one of them to do the talking, preferably your father because, unfairly, men are more likely to take other men seriously. Don't exaggerate, be brief and matter of fact. Then await results.

If you see a child riding a two year old there is not a great deal you can do. Unfortunately, it is not illegal. You can only persuade the child to give up. Tell him or her that it is like expecting a child to work all day in the mines and that the pony's back will be ruined for ever, making him a disabled pony. If the pony is from a riding establishment, complain to the British Horse Society, or the Local Council, and if his condition is bad inform the RSPCA as well.

If you come across a man selling ponies for meat there is little you can do besides buying the ponies because, again, this is not illegal.

If you find an injured rider and he is lying in a heap, send for the ambulance. Don't move him because he may have internal injuries. If the beach is crowded, there may be a doctor among the sun bathers who will help. If he is walking about in riding clothes behaving in a peculiar way, encourage him to sit down and rest while you summon help. Don't offer him brandy, gin or anything to drink.

If you find a dead horse there is nothing much you can do besides letting the police know so that it will be removed.

If your tack is stolen ring up the police and inform them of the theft. You might find someone who could lend you a set for the time being. It is a good idea to insure your tack against theft.

What did Sarah do Wrong?

Here is a cautionary tale. It is a test for you. When you have read it see if you can say what Sarah did wrong.

'We feed them at about nine in the morning. Can you manage that?' asked Simon.

'Yes, no trouble,' replied Sarah, tall and lanky and in jeans and a checked blouson jacket.

'They have a scoop of pony nuts each and a double handful of oats. It's all in the feed store over there. Do you want to have a look?' asked Amanda.

'Is it in a bin?' asked Sarah.

'No, in marked sacks. The sugar-beet cubes are in sacks. Their feed bins are in the paddock. Just take it in a bucket. They are very good, they don't kick or anything, there's really nothing to it,' Simon said.

'And the water needs checking from time to time. You have to turn on a tap because we use a hose. Come and look,' said Amanda.

It was summer. The Andersons' ponies were standing under an oak tree, their coats gleaming, their eyes shining with good health. The children looked at the water trough. It was brimming over with water.

'We check it every day,' said Simon, 'because sometimes birds fall in and drown.'

'We'd better give you the address of the place we're staying at,' suggested Amanda. 'Come inside.'

The house was modern and called 'Moot Point'. Simon wrote an address and telephone number on a piece of paper. 'Okay?' he asked, giving it to Sarah with a smile.

'We hope you don't have to use it,' said Amanda. 'After all we aren't going to be away for long.'

Sarah could see that their car was loaded already. Mrs Anderson was locking the boot and Mr Anderson was letting down the tennis net in the hard tennis court behind the house.

'We'll bring you back a present,' promised Amanda.

'Or would you rather have money?' asked Simon.

'Neither. Don't bother,' replied Sarah, laughing. She left them talking in the drive.

'Everything all right?' asked her mother when she reached home.

'Yes fine. No problems. Just one feed a day at 9 am,' replied Sarah.

'I'll wake you at eight thirty then,' said her mother who rose at seven herself each morning.

* * *

Sarah was late the next morning. Her mother shook her at eight thirty saying, 'Sarah! Wake up. You've got to feed the Anderson ponies, remember?'

She brought her a mug of tea ten minutes later and shook her again. Sarah sat up and looked outside at the sunlight. 'Their ponies won't come to any harm,' she thought. 'Let them wait a bit, it won't kill them.'

Finally at ten o'clock she set off along the road to 'Moot Point' on her bicycle. It was a very old bicycle and the chain came off three times, so when she reached the ponies' paddock she was hot and cross and her hands were covered with black grease. The ponies were waiting for her with eager faces. She threw down her bike and ran to the food store, found the scoop, filled it with grey cubes from the nearest bag, then with oats. She looked them over quickly while they ate, and waited, but not for long, because she suddenly remembered that there was a good programme on television in half an hour. The chain came off again on the way home, and she was in such a rage that she actually kicked her bike, denting a mudguard.

Once home she rushed indoors and switched on the television.

'Everything all right?' asked her mother, peering round the sitting room door.

'Yes, of course. What do you think?' retorted Sarah rudely.

* * *

The next day she was late again. Her father had repaired her bicycle, so she had no trouble with the chain. But the ponies were not at the gate waiting and for one awful moment she thought, 'Oh God, they've escaped!' Then she saw them lying under the oak tree. She fetched their feeds and called but neither moved. She shouted again and one twitched slightly and then lay still. Now her heart was beating in a funny way, and her legs felt as weak as water. She ran across the paddock to the ponies and knelt down beside them. She realized that they were very ill, perhaps even dying. She ran back across the paddock to the house, but everything was locked; she banged on the doors and wrenched at the windows. Then she leapt on her bike and rode full speed for home.

'They're ill!' she screamed, running into the house. 'The Anderson ponies are terribly ill! What shall I do? Oh God, what shall I do?'

While her mother dialled their vet's number she had the feeling that there was no hope, that they really were dying and that it was all her fault.

What did Sarah do wrong?
For answer turn to page 83.

PLACES TO VISIT

SEA PONIES

Galloping close to the sunlit sea,
With wild white horses dancing free,
Over the sandy beaches wide –
Dashing away from the foaming tide.
Freshening gusts in a salty gale,
Stingingly sharp as icy hail,
Fiercely tugging at curls and mane,
Kisses of spray caressing my rein.
Obstacles covered in seaweed green,
Challenging fences scarcely seen . . .
Who shall hold us? Up and away –
Ride to the end of the world to-day!

C. Castellan

How to Find Out About Places

There are many places of interest scattered across the British Isles where you can see different breeds of horses and collections of horse-drawn vehicles. Three are described here in detail. The list of places we include on page 55 does not pretend to be comprehensive. If you want to find out more, consult libraries in the area where you are, or travel agents, hotel receptionists, local papers and of course *Horse and Hound*, all of which should be able to help you. Or you can write to your local Tourist Centre – there are more than 450 in Britain and details and addresses can be obtained from Tourist Information Centre, 64 St James's Street, London SW1.

A Visit to the Royal Mews

The Royal Mews is an easy place to visit because it is near to Victoria Station. Entrance is by turnstile and the cost is very reasonable. We went as tourists and the first exhibit we saw was the enormous gilded State Coach. Drawn by eight horses but only on truly great occasions, the sheer size of it and the lavish decoration are overwhelming. No view of it on television nor photographs can possibly convey the fairy tale magnificence of this coach. Twenty four feet (7.32 metres) long, 8 feet and 3 inches (2.54 metres) wide and 12 feet (3.66 metres) high with the pole 12 feet and 4 inches (3.76 metres) long, and weighing four tons, it is hardly a vehicle for everyday use! It is covered with magnificent decoration. There are panels painted by Giovanni Battista Cipriani on the sides, back and front and three cherubs representing the genii of England, Scotland and Ireland on the roof. It has been used for every coronation since George IV.

After the State Coach we visited the horses. Usually there are thirty horses stabled here but on this occasion many of them were still on holiday at Hampton Court. However, we did see eight horses standing in duck egg green and brown stalls with elegant brass fittings. Bedded on straw, they wore elegant rugs with ER and a crown on them.

The greys were Irish and German bred. The bays Cleveland or Cleveland crossed with Olenburgs with little or no white on them. They seemed a little bored and we had the feeling that they were there just for visiting time, but we may have been wrong.

After the horses we saw more fantastic vehicles – the glass coach, the 1902 State Road Landau, a mail phaeton, sleighs, a miniature barouche and many, many more.

Next we visited the harness room. Here we saw all kinds of harness, some magnificently decorated, some humbler and old fashioned like the flat seated saddles used by past and present members of the Royal family. There were two saddles used by the Queen and her sister when they were children and Queen Victoria's saddle, looking enormous for such a small person. The state harness was dazzling, with blinkers emblazoned with the Royal arms, and hames which were masterpieces of decoration. And then we wandered into a private part of the mews and found a stable just like home with horses everywhere and a lovely smell of hay. The horses had long unpulled manes and there were haynets hanging from rings. But it was private and we were hastily removed by the jeans and T-shirt clad girls who work there.

Then, when we were about to leave, a horse box arrived and presently a magnificent dun stallion with a long dark mane was unloaded. He was followed by two greys: they vanished into the private part of the mews but now it was four o'clock and time for the public to leave. Cameras were put away, the staff in top hats relaxed; the men from Poland, who had bought carriage lamps in Reading, went to look at the exercising vehicles while the rest of us strolled out into the September sunshine, still slightly bemused by so much splendour.

The Royal Mews can be visited on Wednesdays and Thursdays only, from 2 to 4 pm throughout the year except during the Royal Ascot meeting in June, at Bank Holidays or when ceremonial processions are due to take place. Coach parties, etc, must be booked in advance. Entrance in 1980 cost 25p for adults and 10p for children.

For further information see page 56.

But try to visit in autumn or winter when the full complement of horses is likely to be a residence. And be punctual – two hours is hardly long enough to see everything!

Heavy Horse Centres

Until the end of World War I, nearly all heavy, short haul work was done by shire horses. They pulled ploughs and harrows, brewery drays and lumber wagons, in fact great wagons of all kinds. Once these great shire horses were reared mostly in Lincolnshire, Cambridgeshire and Huntingdonshire, but nowadays they come from all over the place and more and more are again being bred. The shire's ancestors probably carried knights in armour and were English horses crossed with the heavy horses of Flanders. They stand around 18.2 hh, that is over six feet or 1.80 metres at the withers, but they are normally gentle and kind.

We visited two shire horse centres, one at Littlewick Green, near Reading, and one at Dedham Vale, near Colchester.

Photographs and rosettes adorn the café at the Vale of Dedham Heavy Horse Centre, eight miles from Colchester. You can buy enormous worn horse shoes for about 10p each, or prints of magnificent horses working, or tea towels with a shire horse printed on them. There is a series of drawings of wagons you can buy – one of these is a Carrier Wagon circa 1710. These wagons, pulled by a pair of shire horses, carried several tons and took six days to complete the journey from Cambridge to London. Their rear wheels were six feet (1.80 metres) high and the wagons fourteen feet

(4.20 metres) long. Considering how bad the roads were in the past you can imagine that the mud and mire must have been terrible.

The Heavy Horse Centre is very much a family business, with the owner, his wife and three daughters involved, and their first pony, Romany, giving young visitors pony rides. The atmosphere is that of a well run private yard.

The show wagons and drays are enormous. The harness room hung with more than twenty sets of harness makes one realize the work involved, and the strength needed to hoist such tack on to the back of an 18.2 hh shire.

The horses had just been brought in for the visitors when we arrived. Most of them were busy polishing off feeds – they are given about 40 lbs (18 kilos) of food a day in summer and up to 70 lbs (32 kilos) during the winter, as well, of course, as grass.

A six month old foal costs between £500 and £1,000 and the owner told us that he likes to buy them at that age, before they are spoilt. It makes breaking in easier. As a four year old shire can cost £2,000 to £4,000 it obviously makes economic sense too.

We were told that all the horses go out every day, even in winter. Probably because of this they can be fed oats (unlike the shires at Littlewick Green who are given not more than twenty pounds a day of dry feed, excluding oats) as well as boiled barley, soaked sugar-beet pulp, bran, chaff and of course hay.

Shires work and are shown in matched pairs.

We saw quite a few four year olds and there were some youngsters at grass elsewhere. Any of these which don't reach the required height are sold on, but only to first class, rigorously inspected homes (one has the feeling that horses come before money in this establishment, and rightly so).

The heavy horses at Dedham Vale have many sponsors – Watney-Manns, Trumans, Anchor Foods, the Eastern Counties Newspapers among them. They can be seen in all their glory performing on many occasions and at many horse shows up and down the country, and of course at the Harness Horse Parade on Easter Monday in Regent's Park, London. They are also to be seen pulling the Lord Mayor's Coach in Norwich on important occasions, just as horses have for so many years.

For less horsey visitors to Dedham Vale there are African Pigmy goats as well as horses, long horned Highland cattle, and Golden Pheasants and peacocks. You can ride in a pony cart pulled by a grey pony, or imagine your forebears riding in the horse drawn vehicles on display. In the harness room, staring at the magnificent team harness, you can relive something which has gone, for everything here is authentic. Even the horse brasses are a little flattened and dented by the years. And, if you feel like crossing a few paddocks and climbing a small hill, you can gaze down across Constable country on one of the most beautiful views in Essex.

There were nine shires at the centre when we visited it, and one retired Suffolk Punch, a beautiful gleaming chestnut. I asked why they kept shires rather than Suffolk Punches, and was told that shires have more show, better action and carriage. In a word I suppose you could call it 'presence'.

Certainly the shires we saw are worth a visit, and there is something for everyone at this centre, even for little brothers and sisters, and grandparents.

The Heavy Horse Centre is open every day in August and on Sundays only from May to October. Prices are very reasonable all season and there is a café and a gift shop. For further details see page 56.

A Visit to Kilverstone Wildlife Park, Norfolk

On entering the park the first animals we saw were the llamas staring at us inquisitively from a paddock on the left.

The owner, Lady Fisher, keeps a number of different species at Kilverstone. At the time, she even had a baby monkey which she wheeled around in a pram.

First we saw the miniature donkeys. Not more than 27 inches (67 cms) high, they are well worth a visit. There were two stallions called Black Jack and Mr Long Ears and three mares, Ass-trid, Piccalino and Rebecca, and everyone was hoping that donkey foals would soon be born.

Next we went to see Blossom, a very small grey Clydesdale mare who is in foal to a miniature horse, so in time there should be a miniature Clydesdale at Kilverstone! She had a little grey Shetland as a companion, and strange neighbours – a herd of Guanacos (wild llamas) from the Andes, who were very chatty in an angry way, the male putting on an extraordinary show of bucks and snorts.

Then came the highlight of our visit, the animals for which Kilverstone is best known, the Falabella horses. Falabellas are miniature horses and true examples of the breed must stand under 34 inches (85 cms) high. They have two less ribs and two less vertebrae than other horses, and the mares can still give birth when they are forty years old, though they take longer to produce their foals, nearly thirteen months instead of the usual eleven. They are a long lived breed, some having been known to live to the ripe old age of fifty!

They don't resemble ponies, but are true little horses with the bones of thoroughbreds. They all come from the Argentine and since Señor Falabella rarely sells stallions we are very lucky to have them in Norfolk. Lady Fisher told us that they have marvellous placid temperaments and that one of the stallions likes to be inside her house as much as possible. They were certainly much more manageable than usual stallions.

At three o'clock it was time for the famous miniature Hickstead, which is held most afternoons between two thirty and three o'clock. A crowd had gathered, pressing against the fence in their efforts to get a better view. The stallions jumped first in hand, that is on a leading rein, because of course Falabellas are too small to ride, but sadly there were no clear rounds. Then Buster the monkey joined in as the budding horseman, and to everyone's delight jumped Crispin successfully.

After the Hickstead we saw a collection of little vehicles suitable for miniature horses to pull, and the old stables, full of shining harness and brass and still used today.

Apart from horses at Kilverstone, there are maned wolves, various monkeys, parrots and other rare birds, jaguars, penguins and rare animals of all sorts. There is a great deal here to appeal to the non-horsey among us. We spent two hours at Kilverstone, but one could spend much longer. There is a picnic area and a café too for people who need constant topping up with fizzy drinks, ice cream, or anything edible. There is a putting area near the entrance for little brothers and sisters, a Latin American Zoo and a gift shop packed full of souvenirs. So if you want to spend hours with the horses, no one need grumble at you.

Open from 10 am to 6.30 pm daily except in winter when it closes at 5 pm. The gift shop and café are open daily throughout the summer season.

Dogs are not allowed inside the park as they might disturb the animals.

There are usually around thirty horses including the Falabellas at Kilverstone, but this rises to thirty-six if you include the donkeys, Blossom and her little Shetland friend Pandora.

Also in East Anglia

While you are in East Anglia, the Horse Rescue Fund is worth a visit. This is run by the Walbanke family, at English Cottage, Great Common, Ilketshall St Andrew, near Beccles, Suffolk, and is a charity which rescues horses and donkeys starving, ill-treated, and in distress or destined for the slaughterhouse, nurses them back to health and then finds them good homes.

We visited it on Open Day (usually held in August) when all the rescued animals come back and parade. It is a small establishment, lovingly run.

We were told that adopting one of their animals is as difficult as adopting a child, and certainly all of them looked happy and in good condition. The Open Day is held to raise funds and is opened by a celebrity. At other times you must telephone before visiting. See page 57 for details.

Also in East Anglia, close to the Shire Horses at Dedham Vale, is the Munnings Collection, and for anyone interested in art as well as in horses this is a must. Alfred Munnings actually lived in Castle House for forty years and the paddocks where he kept his horses are still there. His hunting and racing pictures must be among the finest in the world, and there are many others of more homely cobs and ponies. The Sir Alfred Munnings Art Museum, Castle House, Dedham, Essex, is open only on Sundays and Wednesdays. For details see page 57.

Native Pony Breeds Centres

Dale and Fell Ponies

Years ago when roads were rough or did not exist, nearly everything was carried on the backs of horses, ponies or mules. There were the famous Chapman horses in Yorkshire, who were probably the ancestors of the Cleveland Bay. There were the New Forest ponies who carried smuggled goods inland from the shores of the Solent. There were the Exmoors, the Dartmoors, the Highland ponies, mules and donkeys. They all carried loads on their backs, but probably none worked harder or for longer than the Dale or Fell ponies. Among their loads they carried coal, lead and iron ore from the mines in Durham and Northumberland to the sea. Controlled by one mounted man they followed each other down pack trails through rugged hills, carrying sixteen stone (100 kilos) on their backs. They travelled between two hundred and two hundred and fifty miles in a week with about twenty ponies to a convoy, or 'train' as they were sometimes called. Often they carried their loads right over the Pennine Mountains.

At first the two breeds were very much alike but gradually the Dales became larger by as much as two inches (5 cms) and they now stand at around 14.2 hh. Nowadays they are often used for trekking and are ideal all-round ride and drive ponies. They can be seen at their annual breed society show or through the Dales Pony Society.

Fell ponies stand between 13 and 14 hh. Their Society holds a Stallion Show at Penrith in May, a breed show at Penrith in August and an annual sale at Wigton in October. Some can also be seen living semi-wild on the eastern side of the Lake District.

Shetland

Years ago Shetland ponies worked on the land, carrying or pulling loads of seaweed to fertilize the soil or loads of potatoes and peat in special baskets strapped to their backs. They also carried Scottish crofters and were often the only means of transport. Later they worked in the coal mines, and in Victorian and Edwardian times pulled Bath-chairs at health resorts.

A Shetland should not stand higher than 42 inches. They are very hardy and grow an extra coat in winter to keep out the rain. Sadly many of the foals bred on the Shetland Islands are now probably sent straight to the Continent for meat. However, there are several studs of Shetland ponies both in Britain and overseas. If you want to visit one, or see them in their natural habitat, contact the Shetland Pony Stud Book secretary.

Welsh

There are four types of these. Section A is for ponies up to 12 hh high, Section B is for ponies not exceeding 13.2 hh, Section C is for a cob type of pony not exceeding 13.2hh., and Section D is for the larger - up to 15.2 hh. - Welsh cob. They are all hardy, courageous and good looking. There are records of Arab stallions being turned out with mares in the first half of the nineteenth century.

Welsh ponies can still be seen roaming the Welsh hills. There are herds on the west side of the Brecon Beacons, on the Black Mountains, near Builth Wells and on the Prescelly and Longmynd Mountains, on the Salop Welsh borders and in certain parts of North Wales.

Highland

These ponies vary in height from 13 to 14.2 hh. Well suited to weight carrying, they have quiet temperaments and are often used for pony trekking. Once they lived on at least eleven of the Western Isles. Many were of a lighter breed, particularly on the Isles of Eriskay and Barra where ponies were smaller, 12.2 to 13.2 hh. The ponies of Rhum were mentioned by Dr Johnson and Boswell when touring the islands in the 18th century.

Many Highland ponies have a dorsal stripe and often zebra markings as well. Greys, browns and blacks are common; there are also bays and liver chestnuts and many variations of duns. Small stars on foreheads are allowed, but white socks and too much white on the face are discouraged. There are now studs in England as well as Scotland.

Connemara
These can still be seen in the wilds of Connemara in Southern Ireland. They range in height from 13 to 14.2 hh. Greys are common, but there are also blacks, bays, browns and duns and occasionally roans and chestnuts. They are excellent jumpers with good conformation.

These ponies have lived in Connemara from early times and have been crossed with many breeds including Arab, Welsh, thoroughbreds, half breds, hackneys and Clydesdales. Only in the 1920s when the Connemara Pony Breeders Society was formed did the breed develop its present-day identity and stud book.

New Forest
These vary in size from 12 to 14.2 hh., though there are few higher than 13 hh. in the Forest. All colours are acceptable except piebald and skewbald. They have been crossed with many breeds including Arab, Highland, Dartmoor, Exmoor and even Hackney. They appear frequently in the book *Children of the New Forest* which was first published in 1847 and describes the life in the Forest in 1647 when Cromwell was in power. They are probably the most accessible of all our native breeds, being easy to see at all times of the year in the Forest and close to roads passing through.

Dartmoor

The maximum height for these ponies is 12.2 hh. The most common colours are black, bay and brown. Piebalds and skewbalds are banned. Arabs, thoroughbreds and other breeds were turned out on the moors with them, followed by Shetlands at the time of the industrial revolution when pit ponies were in demand. In 1899 there was a successful attempt to improve the breed and today's Dartmoor is full of quality with a good shoulder and excellent conformation.

These ponies can be seen grazing on Dartmoor.

Exmoor

These ponies are easy to recognize. Sturdy with large eyes, they have distinctly mealy coloured noses. Their maximum height is 12.3 hh. and they have wide chests and are often miniature weight carriers. Bay, brown or dun are acceptable colours. They are mentioned in the Domesday Book and probably arrived in Britain during the bronze age.

Like most native breeds they are good doers and have great stamina. They can still be seen on Exmoor and until 1830 were used as pack ponies. They were also used in coal mines and in harness by farmers. Shepherds rode them for herding sheep.

If you wish to know more about these native breeds and where you can visit them, write to the individual breed's Society secretary. A list of their names and addresses is given on the following pages.

Addresses of Places to Visit

JOUSTING TOURNAMENTS

Belvoir Castle, Grantham, Lincolnshire (also the home of the military museum of 17/21st Lancers). Tel Knipton 262 or 233.

Chilham Castle, near Canterbury. Tel (022 776) 319.

Knebworth House, near Luton, Bedfordshire. Tel (0438) 812661.

UNUSUAL BREEDS

Prewalski Wild Horses. Port Lympne Zoo Park and Gardens, Lympne, Hythe, Kent. Whipsnade Zoo, Dunstable, Bedfordshire.

Mongolian Wild Horses. Marwell Wild Life Adventure-Land, Golden Common, near Winchester, Hampshire.

Falabellas – Argentinian miniature horses. Kilverstone, Norfolk. (Open daily 10 am to 6.30 pm, 5 pm in winter.)

Shires. Dedham Vale, Colchester, Essex. (Open daily in August, otherwise Sundays only May to October.) Littlewick Green, near Reading, Berkshire. (Open throughout the summer.)

NATIVE BREEDS

Dale. Mr C. H. Hodgson, Ivy House Farm, Hilton, Yarm-on-Trees, North Yorkshire.

Fell. Miss P. Crossland, The Packway, Windermere, Cumbria.

Shetland. Mr D. M. Patterson, 8 Whinfield Road, Montrose, Angus, Scotland.

Welsh. Mr T. E. Roberts, 6 Chalybeate Street, Aberystwyth, Dyfed, Wales.

Highland. Mrs J. M. Thomson, Western Sunnyside, near Mothven, Perth, Scotland.

Connemara. Miss Sarah Hodgkins, Buttermilk Farm, Leafield, Oxford.

New Forest. Miss D. Macnair, Beacon Corner, Burley, Ringwood, Hampshire.

Dartmoor. Mrs E. C. M. Williamson, Weston Manor, Corscombe, Dorset.

Exmoor. Mrs J. Watts, Quarry Cottage, Sampford Brett, Williton, Somerset.

HORSE-DRAWN VEHICLES

Royal Mews, Buckingham Palace, London. State coach and other ceremonial vehicles.

The London Transport Museum, Covent Garden (there are other transport museums scattered across Great Britain).

Tyrwhitt Museum of Carriages, near Maidstone, Kent. Largest display of horse-drawn vehicles in England.

Staffordshire County Museum and Mansion House, Shugborough, Staffordshire.

Breamore House, near Fordingbridge, Hampshire.

Gretna Green, Scotland.

Dodington House, Chipping Sodbury, Avon. Over 30 vehicles plus carriage horses.

Bath Carriage Museum, Bath, Avon. Thirty vehicles.

OTHER PLACES OF INTEREST

Horse Rescue Fund, English Cottage, Great Common, Ilketshall St Andrew, Beccles, Suffolk. Tel Ilketshall St Andrew 295.

Donkey Sanctuary, Woodley, near Reading, Berkshire.

Munnings Collection, Art Museum, Castle House, Dedham, Essex. Fine collection of paintings by Sir Alfred Munnings. Tel Colchester 322127.

Horse Museum, Belton House Park and Gardens, Grantham, Lincolnshire.

Zoological Park, Murrayfield, Edinburgh, Scotland. Zebras and ponies.

Cardiff Zoo, Weybeck Road, Barry. Shetland ponies and donkeys. Open April to October.

Chessington Zoo, Chessington, Surrey. Zebras.

Arlington Court, near Barnstable, Devon. Herd of Shetland ponies. Reputed finest collection of horse-drawn carriages in country. Rides in a coach and pair, except Mondays. Open daily 1 April to 31 October.

Trekking Centres

There are many centres, good, bad and indifferent. *Ponies of Britain* issues a yearly list of approved ones where the ponies are in good condition and the centre well run. So if you plan to trek begin by writing to *Ponies of Britain*, Brookside Farm, Ascot, Berkshire, enclosing a stamped addressed envelope. They will reply, telling you the current cost of their information brochures. Their list will include riding holidays too and post trekking centres, and will give you an idea of what each centre provides as well as how much it will cost.

Scotland
Erray Estate, Tobermoray, Isle of Mull, Argyll. Tel (0688) 2052. Ormsaig, Ross of Mull. Tel Pennyghael 230.

Wales
Mid-Wales Riding Centre, Llanwrtyd Wells. Tel Llanwrtyd Wells 300.

Dartmoor
Shilstone Rocks Dartmoor Stud, Widecombe-in-the-Moor, Newton Abbot, Devon. Tel Widecombe 281.

Exmoor
North Wheddon Farm, Wheddon Cross, Minehead, Somerset. Tel Timberscoombe 224.

New Forest
Decoy Trekking Centre, Lyndhurst, Hampshire. Tel (042 129) 2652.

Lake District
Robin Hood Riding Centre, Bassenthwaite, Keswick, Cumbria. Tel Lake 296.

Yorkshire
Swinton Riding and Trekking Centre, Home Farm, Masham, near Ripon, North Yorkshire. Tel. (0723 85) 218.

The 1980 Description of Trekking and Riding Holiday Centres Approved by the Ponies of Britain contained more than 130 centres in England, 34 in Scotland, 46 in Wales, 2 in Ireland, 1 in Spain, 1 in USA, 1 on the Isle of Man and 12 Post Trekking, Trail Riding or Tours Establishments, so you should find one to suit somewhere!

Annual Events

Badminton Horse Trials, Gloucestershire *April*
Hickstead Showjumping, Sussex *April*
Oxford County Show *April*
Harness Horse Display Regent's Park *Easter Monday*
Royal Windsor Show *May*
Newark and Notts Show *May*
Devon County Show *May*
Shropshire and West Midland Show *May*
Tidworth Horse Trials, Wiltshire *May*
Suffolk County Show *end of May, beginning of June*

Bath and West Show *end of May, beginning of June*
Royal Highland Show *June*
Bramham Park Horse Trials, Yorkshire *June*
South of England Show *June*
Royal Norfolk Show *June*
Great Yorkshire Show *July*
Hickstead, Sussex *July*
Royal International Show, London *July*
Greater London Show, Clapham Common, London *August Bank Holiday*
Royal Lancashire Show *August*
Hickstead *August and September*
Burghley Horse Trials, Lincolnshire *September*
World Driving Championships, Windsor, Berkshire *September*
Wylye Horse Trials, Wiltshire *October*
Horse of the Year Show, Wembley, London *October*

RED

RAINY DAY MISCELLANY

LIMERICK

There once was a boy called Ted,
Who wanted his pony to be red,
He painted him all over,
The colour of clover,
  But he cried,
  When he died,
  From the lead

Pony Feats

In stage coach days, a Fell pony and a stage coach left Exeter at the same time and, though the stage coach horses were changed fifteen times on the way, the Fell pony still reached London first.

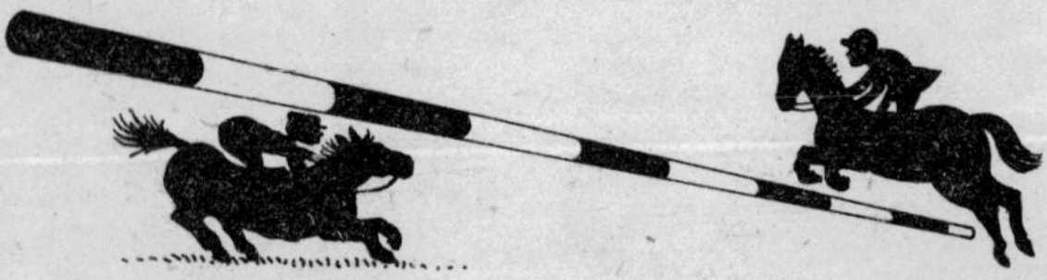

Sambo, an 11.3 hh. Welsh pony, became champion jumper at Olympia in 1887. It was said that his rider could take him under a bar and then turn him round and jump him over it!

A pair of Shetland ponies only 37½ inches (93 cms) high travelled eleven miles (17 km) in sixty minutes pulling three adults in a four wheeled vehicle.

A Shetland mare in Yorkshire trotted a mile (1.6 km) in three minutes forty four seconds driven by a man weighing more than eleven stone (70 kilos).

A pony stallion owned by a farmer and living in Central France was reputed to be fifty-four years old in 1919 and in June 1970 a Welsh pony, Chess, living on a farm in West Glamorgan was said to be sixty-six years old.

Games and Puzzles to do While Travelling

I spy, a game for one or two players

For 1 player
See how quickly you can spot and tick off all the objects on the list. Either tick them off as you see them or in order.

For 2 players
Start at opposite ends each with a chosen playing counter. When you spot the sighting in the first square (eg, a horsebox or trailer, or donkey) you move your counter forward. Repeat this for every sighting until one of you finishes.

START FINISH
A horsebox or trailer
A riding school, or a sign indicating one
A dun horse or pony
Bales of hay or straw
A pub sign with a horse's head
Saddler's shop
Horse shoe
A skewbald horse or pony
A piebald horse or pony
Saddle and bridle
A jumping fence
A gate, open or closed
Two mounted horses or ponies
A strawberry roan horse or pony
A barn
A donkey
START FINISH

Anagrams

Unscramble the authors' names and their books and pair off the authors with the books they've written.

1 BURY REGUFONS	A TEH ERD YONP
2 CAPIRAIT CLIEHT	B NOPY TORALP
3 NERCIOAL RIVLES	C HIGNT FO HET DER SHORE
4 TICHRIENS NELPIUL POTHNOMS	D MUMERS TWIH MOMTY
5 HIJUDT SORERKSFIB	E A TABLES ROF ILJL
6 HOJN NECKSTEIB	F KAJCEI NAD EHT TISMIF NOYP

1+E, 2+C, 3+D, 4+B, 5+F, 6+A

For answers turn to page 83

Codes

Can you decipher the following horse and pony events by using one of the codes below?

eg, CADMHOUNO INRSF UQHAKS = BADMINTON HORSE TRIALS Using the 5th code

1 UBEFR BS GUR LRNE FUBJ

2 HSBOE OBUJPOBM

3 GSV WVIYB

4 PQHOBF PIHKHP MNTOUFD GAMFS
5 AZCLHMSNM GNQRD SQHZKR
6 SRXPHGVZW

For answers turn to page 83

1 Alphabet split
eg, PBQR = CODE

A B C D E F G H I J K L M

N O P Q R S T U V W X Y Z

2 *Alphabet reverse*
eg, XLWV = CODE
A B C D E F G H I J K L M N O P Q R S T U V W X Y Z
Z Y X W V U T S R Q P O N M L K J I H G F E D C B A

3 *The 'letter before' code* eg, BNCD = CODE
4 *The 'letter after' code* eg, DPEF = CODE
B C D E F G H I J K L M N O P Q R S T U V W X Y Z A
A B C D E F G H I J K L M N O P Q R S T U V W X Y Z

5 *Two up and one down*
eg, BNDF = CODE

A B D E G H J K M N P Q S T V W Y Z
A C D F G I J L M O P R S U V X Y

Crossword 1

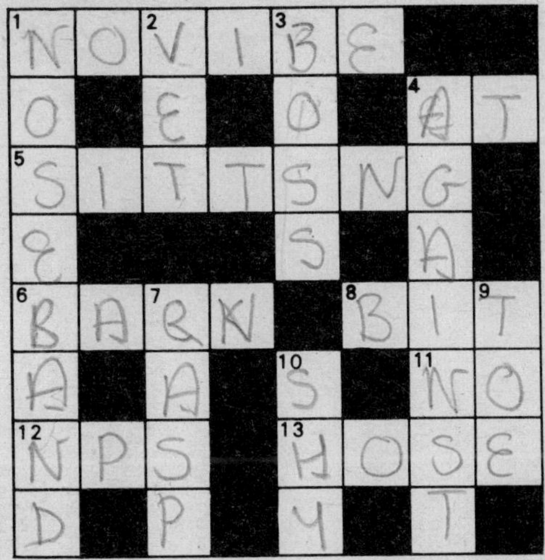

Across

1 An inexperienced horse or rider (6)
4 A horse may be - - grass (2)
5 One way of trotting (7)
6 Where we might keep the hay (4)
8 The hackamore does not have this (3)
11 Negative (2)
12 You are probably interested in this organization (3)
13 This should be used on a swelling on your pony's leg (4)

Down

1 A part of the bridle (8)
2 A doctor of horses (3)
3 The manger (4)
4 - - - - - - - the clock (7)
7 We do this to a horse's teeth (4)
9 The chestnut is all that remains of this prehistoric feature (3)
10 Your pony may do this if he is afraid of something (3)

Turn to page 84 for the solution

Maze

Help Nicky to find her pony

For answer turn to page 84

Word Search 1

Breeds and colours

```
A E A Q I P D U N O F H S L L E F P L T
R S T R P R U V W E K A H O L L O I R C
U D H R A I S T L C N C E S O T N I P L
N L A I K B I R A E U K T P O C G R E Y
X A L O R A O L S H E T L A N D O M Y A
E B I N E E B O P A L O M I N O N R T S
B W B V I A T K A H C O N N E M A R A O
A E N A O R Y R R E B W A R T S D A L O
Y K L O A C I P Z O H D I A W A E R P L
U S N A O R E U L B L S F O L R U S S A
T U N T S E H C H A C P L O S B E I Z P
D R B O Y U L U B S O I T E R K I L A P
A P N O O F C E B R O W N M W L C N A A
K I G H C U I B M U S T A N G D O C O D
S O P Y L P Y E R G E L P P A D L V O Q
```

Find as many of the words listed below as you can. You may go forwards, backwards, up, down or diagonally up or down.

Albino
Appaloosa
Arab
Bay
Black
Blue roan
Brown
Chestnut
Cob
Connemara
Criollo
Dale
Dapple grey
Dun
Fell
Grey
Hack
Hucul
Mustang
Orlov
Palomino
Piebald
Pinto
Russ
Shetland
Shire
Skewbald
Strawberry roan
Viatka
Welsh
cod

For answer turn to page 85

More Anagrams

Unscramble the races below

1 GINBEND
2 PONDA GON GES
3 POCE REAR
4 TREW TALK CARNOT
5 TOT ACE ROAP
6 SUM SCALE POLI
7 GLACE RAF
8 NUN DEAL RAD

For answers turn to page 85

Brain Teasers

Seven remaining horses in a race

If RUM was behind DUKE, who came in third, SAPPHIRE came in behind JEWEL, who was before CLOUD, who was behind RUM, who was in front of SLUMBER, and RUM was between CLOUD and BELLE, who were both behind SAPPHIRE, who was in front of DUKE and SLUMBER, who was also behind CLOUD –

who won the race and what was the order of the other horses finishing?

For answers turn to page 85

'Where' Quiz

1 Where is the Spanish Riding School?

A Bonn, Germany
B Vienna, Austria ✓
C Madrid, Spain

2 Where is the International Horse Show held every year?

A Hickstead
B Stoneleigh
C Wembley ✓

3 Where were the Olympic Games held in 1968?

A Italy
B Mexico ✓
C Canada

4 Where used the Horse of the Year Show to be held?

A Harringay Arena ✓
B Reading Stadium
C Windsor Great Park

5 Where is the Cadre Noir?

A Belgium
B Australia
C France

6 Where is the National Stud?

A Ledbury
B Taunton
C Newmarket

7 From where does the breed of horse 'Nonius' originate?

A Hungary
B Yugoslavia
C Germany

8 Where is the Grand National held every year?

A Kempton
B Ascot
C Aintree

9 Where are the Royal Mews?

A Buckingham Palace
B Windsor Castle
C Sandringham House

10 Where are the Prince Philip Cup finals held?

A Skegness
B Wembley
C Stoneleigh

For answers turn to page 86

Crossword 2

Across

1 This may be done sitting or rising (8)

5 A game played on horses (4)

6 - - - - - and rails (5)

8 An extra hunting day not previously advertised - - - day (3)

10 A horny growth at the back of the fetlock joint (5)

12 A rider does this to a fox when he causes it to change direction (5)

Down

1 A type of jump (6)

2 Food for horses (4)

3 Shoes used to protect the feet of horses out at grass (4)

4 - - - - - rumped (5)

7 This may be rasped (5)

9 The place outside the stables (4)

11 This may be a New Zealand one (3)

Turn to page 86 for the solution

Step by Step

By solving the twelve clues and writing them across you'll find nine more four letter words reading vertically.

1 Opposite of wild
2 A sturdy breed of pony
3 A colour
4 What a saddle can cause if used wrongly
5 Front leg
6 Back leg
7 Newmarket is famous for its Thoroughbred - - - -
8 Mature female
9 Hay and straw come in this
10 An obstacle at Hickstead
11 A stable should not be - - - -
12 Wrong fitting shoes can cause a horse to become - - - -

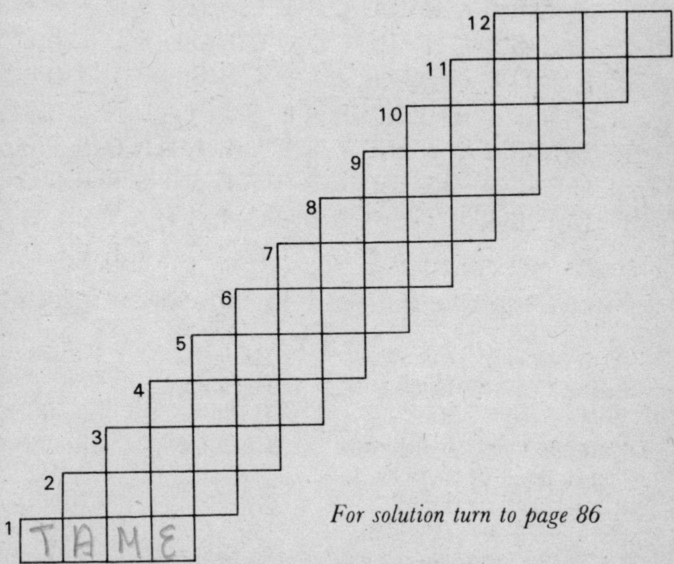

For solution turn to page 86

Word Search 2

Points of a horse

C A N O S T R I L S T I S N I O L S A E
H O O S T P A S T R N P O B A S S O L P
E A R E S O O K R H I G H S M I C T N U
S E F E T L O C K Z R A R E S J O W L O
E D O D H E A L J O W O E S E E N K O R
R E D L U O H S B R E T A M U I L O O C
L I P K C O L E R O F L A T I D A N S O
E A R S A C A N N O N B O N E T O L P E
B A C T V I L E C F O O T S O M E C A S
C O R O N E T H E T F O R E A R M I K S
E D O N R E E A F E V O O R G N I H C T
T A K E S E T O O P S E G S F O O H F I
P A R R K H L U C R F L A N K P M A R F
S U S B E T H E E D C O H O A P O L L L
K E O R X O M H A L L T H S I R T Y O E
A N I F C L T H R A N E T H A U N C H T
E N R K S I H E E L S E P A N N O S T R
G R O A W E B O N S R T U N T S E H C O
N I B R E A S T A N P O T S H A Y E L A
M R A F E T S N O L T O T B A R R E L T

Find as many of the words listed below as you can

Arm
Barrel
Breast
Cannon bone
Cheek bone
Chestnut
Chin groove
Coronet
Croup
Dock
Ears
Feathering
Fetlock
Flank
Forearm
Forelock
Haunch
Heels
Hock
Hoofs
Jowl
Knee
Loins
Nostrils
Pastern
Poll
Shoulder
Stifle
Throat
Withers

For answers turn to page 87

Crossword 3

Across

1 We win these in competitions (8)
5 A gait (6)
7 These may be attached to mouthing bits (4)
8 A point of the horse (6)
10 The jackass and the - - - ass (3)
12 These are used in betting (4)

Down

1 The turf (6)
2 Sometimes a groom will - - - - - a horse's hair when he is clipped (5)
3 A cross-country holiday ride (4)
4 This type of boot is strapped round the coronet to prevent a capped elbow (7)
6 A part of the bridle (5)
9 A signal from the rider to the horse (3)
11 'hands high' (2)

Turn to page 87 for the solution

Word Ladders

In a word ladder, each word is changed into the next by changing just one letter. In this example FORE becomes SORE by changing the F only, and SORE becomes SIRE by changing the O into an I.

FORE
SORE
SIRE

Here are two word ladders for you to complete. Can you change BEAT into FEED in only three steps and DALE into FELL in four steps?

BEAT
FEED

DALE
FELL

For solution turn to page 88

ANSWERS

What did Sarah do wrong?

Sarah was lazy and always in a hurry. She arrived late to feed the ponies, did not check their water and was only interested in returning home as soon as possible. But her one disastrous mistake was to rush into the feed store and scoop up the first cubes she saw in the sack! These were sugar-beet cubes, which must be soaked for twelve hours before being fed to ponies. Had she taken more trouble, she would have seen pony cubes, which can be safely fed dry, in another sack. The sugar-beet cubes swelled inside the ponies' stomachs causing blockages followed by acute colic. Had she arrived earlier the next morning, she might have alerted a vet in time; as it was both ponies died.

Anagrams

1	Ruby Ferguson	E	*A Stable for Jill*
2	Patricia Leitch	C	*Night of the Red Horse*
3	Caroline Silver	D	*Summer with Tommy*
4	Christine Pullein-Thompson.	B	*Pony Patrol*
5	Judith Berrisford	F	*Jackie and the Misfit Pony*
6	John Steinbeck	A	*The Red Pony*

Codes

1 Horse of the Year Show (1st code)
2 Grand National (4th code)
3 The Derby (2nd code)
4 Prince Philip Mounted Games (5th code)
5 Badminton Horse Trials (3rd code)
6 Hickstead (2nd code)

Crossword 1

Across

1 Novice   4 At   5 Sitting   6 Barn   8 Bit   12 NPS   13 Hose

Down

1 Noseband   2 Vet   3 Crib   4 Against   7 Rasp   9 Toe   10 Shy

Maze

Word Search 1

A	E	A	Q	I	P	D	U	N	O	F	H	S	L	L	E	F	P	L	T
R	S	T	R	P	R	U	V	W	E	K	A	H	O	L	L	O	I	R	C
U	D	H	R	A	I	S	T	L	C	N	C	E	S	O	T	N	I	P	L
N	L	A	I	K	B	I	R	A	E	U	K	T	P	O	C	G	R	E	Y
X	A	L	O	R	A	O	L	S	H	E	T	L	A	N	D	O	M	Y	A
E	B	I	N	E	E	B	O	P	A	L	O	M	I	N	O	N	R	T	S
B	W	B	V	I	A	T	K	A	H	C	O	N	N	E	M	A	R	A	O
A	E	N	A	O	R	Y	R	R	E	B	W	A	R	T	S	D	A	L	O
Y	K	L	O	A	C	I	P	Z	O	H	D	I	A	W	A	E	R	P	L
U	S	N	A	O	R	E	U	L	B	L	S	F	O	L	R	U	S	S	A
T	U	N	T	S	E	H	C	H	A	C	P	L	O	S	B	E	I	Z	P
D	R	B	O	Y	U	L	U	B	S	O	I	T	E	R	K	I	L	A	P
A	P	N	O	O	F	C	E	B	R	O	W	N	M	W	L	C	N	A	A
K	I	G	H	C	U	I	B	M	U	S	T	A	N	G	D	O	C	O	D
S	O	P	Y	L	P	Y	E	R	G	E	L	P	P	A	D	L	V	O	Q

More Anagrams

1 Bending
2 Egg and Spoon
3 Rope Race
4 Walk Trot Canter
5 Potato Race
6 Musical Poles
7 Flag Race
8 Run and Lead

Brain Teasers

Jewel won the race. The following order was Sapphire, Duke, Belle, Rum, Cloud, Slumber.

'Where' Quiz

1 B Vienna, Austria
2 C Wembley
3 B Mexico
4 A Harringay Arena
5 C France
6 C Newmarket
7 A Hungary
8 C Aintree
9 A Buckingham Palace
10 B Wembley

Crossword 2

Across
1 Trotting 5 Polo 6 Posts 8 Bye 10 Ergot 12 Heads

Down
1 Triple 2 Oats 3 Tips 4 Goose 7 Tooth 9 Yard 11 Rugs

Illust. Step by Step B — use question artwork combined with overlay

Step by Step

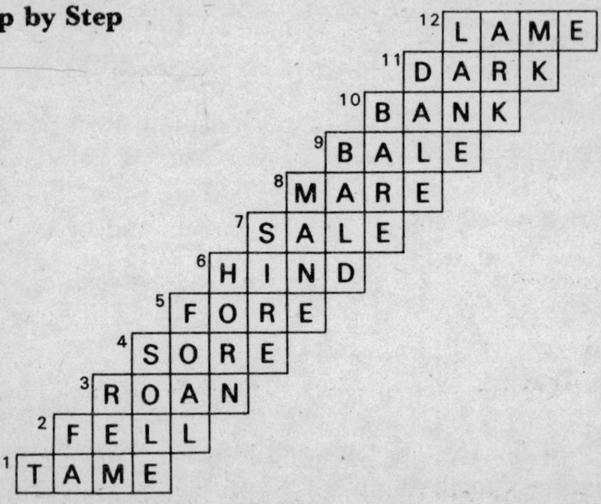

Word Search 2

C A N O S T R I L S T I S N I O L S A E
H O O S T P A S T R N P O B A S S O L P
E A R E S O O K R H I G H S M I C T N U
S E F E T L O C K Z R A R E S J O W L O
E D O D H E A L J O W O E S E E N K O R
R E D L U O H S B R E T A M U I L O O C
L I P K C O L E R O F L A T I D A N S O
E A R S A C A N N O N B O N E T O L P E
B A C T V I L E C F O O T S O M E C A S
C O R O N E T H E T F O R E A R M I K S
E D O N R E E A F E V O O R G N I H C T
T A K E S E T O O P S E G S F O O H F I
P A R R K H L U C R F L A N K P M A R F
S U S B E T H E E D C O H O A P O L L L
K E O R X O M H A L L T H S I R T Y O E
A N I F C L T H R A N E T H A U N C H T
E N R K S I H E E L S E P A N N O S T R
G R O A W E B O N S R T U N T S E H C O
N I B R E A S T A N P O T S H A Y E L A
M R A F E T S N O L T O T B A R R E L T

Crossword 3

Across

1 Rosettes 5 Canter 7 Keys 8 Gaskin 10 She 12 Odds

Down

1 Racing 2 Singe 3 Trek 4 Sausage 6 Reins 9 Aid 11 hh

BEAT
BEET
FEET
FEED

DALE
PALE
PALL
FALL
FELL

Books to Read

To read before you go:
Where to Ride, a list of Approved, Recognized and Listed Riding Schools £1.70
Pony Trekking Glenda Spooner £1.65

And to make you laugh!
Thelwell's Annual £2.25
Thelwell's Gymkhana £2.90
Thelwell's Pony Painting Book 80p

All from the British Horse Society Book and Gift Shop, The British Equestrian Centre, Kenilworth, Warwickshire. Tel: Coventry (SDT 0203) 52241. Prices only approximate.

Horse and Hound will help keep you informed of what is happening and where. And *Pony* will make excellent bedtime reading. A Horse Dictionary and/or Encyclopaedia will help you to cope with quiz and puzzle books and will also teach you a good deal. You can also use it to test the knowledge of your family.

Armada Horse and Pony Quiz Books will give you something to do when your car breaks down or rain falls from the sky in bucketsful.

Adventure Trail, Granada, 85p, gives handy hints on survival; it tells you how to signal for help, how to make drinking water, how to read a map, how to deal with bites and even how to catch fish. It also tells you how to find your way by the stars. Surely a must for all intrepid holiday makers, horsey or otherwise.

Pony Books from Granada.

Mary O'Hara		
My Friend Flicka Part 1	85p	☐
My Friend Flicka Part 2	60p	☐
Green Grass of Wyoming 1	60p	☐
Green Grass of Wyoming 2	60p	☐
Green Grass of Wyoming 3	70p	☐
Thunderhead 1	60p	☐
Thunderhead 2	60p	☐
Thunderhead 3	60p	☐
Christine Pullein-Thompson		
For Want of a Saddle	50p	☐
Impossible Horse	60p	☐
The Second Mount	60p	☐
The Empty Field	50p	☐
Three to Ride	60p	☐
The Pony Dopers	75p	☐
Pony Patrol	40p	☐
Pony Patrol S.O.S.	40p	☐
Pony Patrol Fights Back	50p	☐
Pony Patrol and the Mystery Horse	75p	☐
Pony Parade	60p	☐
Riders on the March	60p	☐
They Rode to Victory	60p	☐
The Open Gate	60p	☐
The Lost Pony	50p	☐
The First Rosette	60p	☐
Anne Digby		
Quicksilver Horse	75p	☐
A Horse Called September	60p	☐